J. Wilson

The Family Companion and Physician

On Reformed Botanical Principles, Treating of the Symptoms and

Remedies of Acute Diseases in Men, Women and Children

J. Wilson

The Family Companion and Physician
On Reformed Botanical Principles, Treating of the Symptoms and Remedies of Acute Diseases in Men, Women and Children

ISBN/EAN: 9783337296995

Printed in Europe, USA, Canada, Australia, Japan

Cover: Foto ©berggeist007 / pixelio.de

More available books at **www.hansebooks.com**

THE

FAMILY COMPANION

AND

PHYSICIAN.

ON REFORMED BOTANICAL PRINCIPLES,

TREATING OF THE

SYMPTOMS AND REMEDIES OF ACUTE DISEASES, IN
MEN, WOMEN, AND CHILDREN;

TOGETHER WITH A

SHORT LECTURE

ON THE LAWS OF THE MANY NERVOUS DISEASES TO
WHICH WE ARE SUBJECT.

ALSO,

NUMEROUS PREPARATIONS,

VALUABLE TO EVERY FAMILY.

By DR. J. WILSON,

FELLOW OF THE ECLECTIC MEDICAL SOCIETY OF NEW YORK STATE,
COOPER INSTITUTE.

OWEGO, NEW YORK, 1866.

21st EDITION.

ROCHESTER:
DAILY DEMOCRAT STEAM PRINTING HOUSE,
62 BUFFALO STREET.

INTRODUCTION.

In presenting this little work to the public, I am confident that it will meet the approbation and approval of the masses. The present age may be emphatically regarded as an age of investigation and improvement, and when truth is fairly presented to the human mind, it seldom fails to meet with the most cordial approval and reception.

In arts and sciences in general, greater researches and discoveries have been made during the last few years than at any period previous—such as steamboats, railroads, education, and various other matters ; the human mind has achieved wonders, and given ample proof of its divine origin.

But in regard to medical treatment, our race may justly be said to have retrograded rather than improved. The common people have had their

minds blurred and biassed by technicalities and foreign languages, as well as imposition, until they
have but little knowledge of the human organiza-
tion, and much less of the simple remedies found in
the wild garden of Nature, for the curing of disease.
And physicians themselves, even the most learned
and skillful, have, instead of improving, greatly retrograded, and now constitute a most lamentable
contrast to the progress made in other departments.
However, light is breaking forth, the day begins to
dawn. Men are investigating these matters. Reform medical schools are established. Physicians
are springing up everywhere. Dagon's god begins
to tremble. King Calomel is tottering upon his
throne, and before long will fall, I trust to rise no
more—Its old rubbish removed, as well as the build-
ing torn down, and a new edifice which is already
going up, and with a few more efforts will be com-
pleted. We may therefore say, with the great po-
litical reformer, Thomas Jefferson, we are in the full
tide of successful experiment, and I would add, improvement.

This little work may justly be regarded as a Family Companion and Physician, and the warmest

friend by whom you have ever been greeted. If you will listen to its counsels it will save you from as well as remove the many pains and aches incident to life.

My object has been in this work to teach you how to cure your acute diseases without the use of calomel, and I am confident, if I can succeed in this, that most of the chronic diseases will be avoided, as it is from acute that nearly all chronic diseases originate. I have endeavored to render the practice plain and simple, and have prescribed such remedies as are attainable by all, so that the common people as well as the learned may cure themselves with equal facility. I have also described the symptoms of each disease in as brief and simple form as possible, so that at a glance you may readily detect your disease and the remedies indicated.

Praying and trusting by the grace of God, that by the aid of this little work our race may be freed, and to prevent many of the ills of life, as well as lengthen their days.

I am, yours, &c.,

THE AUTHOR.

DR. WILSON'S

FAMILY COMPANION

AND PHYSICIAN.

Ague and Fever, or Intermittent Fever.

Symptoms.—This disease may be divided into three classes, viz. :

1. The cold stage, the hot stage, and the sweating stage. The cold stage generally begins with pain in the head and loins, weariness of the limbs, coldness of the extremities, stretching, yawning, and sometimes vomiting, shivering and violent shaking.

2. A burning heat, red skin, sensitive to the touch, pain in the head, and flying pains are felt in the different parts of the body, and often flighty ; the pulse is quick and

strong, the tongue white, the thirst great, and the urine is highly colored.

3. Perspiration is seen first about the neck and breast, and thus continuing until profuse sweat breaks out at every part of the body; the heart diminishes to its usual standard, and the pulsations to their usual number, and all the functions are restored to their natural order—when after a certain interval the paroxysm returns, and produces the same distressing sensations, generally once in twenty-four hours.

To Cure.—First to put as speedy a stop as possible to the fits when they have occurred, give the patient an emetic, viz. : Take equal parts pulv. Lobelia, Ipecac, and Bloodroot, pulv. cayenne pepper one half a tea-spoonful, and repeat every twenty minutes, together with Boneset or Chamomile tea, until the patient vomits freely. This is to be given before the cold stage, and often it breaks up the disease without further medicines. It sometimes happens that the Cayenne cannot be taken ; in that case give less doses, sufficient to vomit a little.

Should any peculiar temperament, debility, or state of the system render an emetic injudicious, you will give a cathartic, viz. : Mandrake or May apple, and Cream-tartar, in equal parts, one tea-spoonful ; add one gill of boiling water ; sweeten to taste, and drink. This will cleanse both the stomach and bowels.

But should the cold stage come on, then you will give the patient warm Catnip or Virginia Snakeroot, Boneset, or Peppermint teas, and apply warm bricks to his feet, cover warm, and use every means to promote perspiration. These remedies will greatly lessen the other stages.

Hot stage.—As soon as the cold stage terminates, you will remove the extra covering, and all warm applications, as well as warm drinks, and give the patient cold drinks. You may also give lemonade freely. This course will allay the febrile excitement.

Sweating stage.—At this stage both cold and hot drinks must be discontinued, and those that are tepid given.

During the intermission drink ague wine

bitters. This will cure in a short time where everything else fails. Drink from a half to one wine-glassful every two or three hours, made as follows : One oz. Peruvian bark, half oz. Wild Cherry bark, half oz. Cinna mon, half oz. Cloves, half oz. Nutmeg, .half oz. Cayenne pepper, one tea-spoonful Sulphur, and one quart Wine. Pulverize all.

Cathartic.—Extract the substance from bark of butternut root, and take one pill the size of a pea each night. These remedies will not fail to cure in a short time.

Remittent Fever.

This disease is characterized by frequent paroxysms, one succeeding another so rapid, that the former scarcely leaves until they have another symptom much like the former, it being very difficult to discriminate between the two, only as the former attacks with the paroxysm but once in about twenty-four hours.

Remedies.—First, give an emetic at once,

combined with the vapor bath, and repeat if necessary the following day.

Second, cleanse the bowels with the Mandrake compound.

Third, give the Quieting powder every two or three hours, and bathe frequently the entire body with warm weak lye or warm saleratus water, until you promote perspiration. You may also give the patient freely warm Lemon tea. Give the patient Slippery Elm, Comfrey root, Mint, or Catnip tea.

Often the patient has severe headache. In that case make cold applications to the head and warm to the feet.

After the fever is removed, the patient will be weak. In that case you will give him the Wine bitters, a half to one wine-glassful three times a day.

Inflammatory Fever.

This disease may be known by great heat, frequent, strong, and hard pulse, redness of the face, and may be distinguished from the former by the increased amount of inflammation, dizziness of the head, &c.

Remedies.—This fever should in every way be treated as the remittent fever, omitting the emetics. First give the Mandrake compound once in four hours, until you promote a healthy action of the bowels, and then follow up the remedies as directed under that head.

Regimen.—Nutritious liquids should be given, such as corn meal gruel, toasted bread and water, ripe fruits, roasted fruit, and water, &c.

All kinds of fever may be broken in a short time by the use of the emetic or mandrake compound and vapor bath, if taken in time.

Continued Fever.

This fever is characterized by debility, inactivity, heaviness, yawning, stretching, coldness in the back, which continues to spread until it sweeps over the whole system, accompanied with chills, &c.; stomach nauseated, and frequently confusion of intellect. Finally the coldness passes off, and then the skin becomes dry, face red, dull pain in the head, the pulse quick and full, great tendency of blood to the head, costiveness, urine highly colored and scanty.

Remedies.—Follow the directions as laid down under the head of Remittent and Inflammatory Fevers. At the beginning give the Emetic, and in eight or ten hours after give the Mandrake compound, and repeat every four hours until you obtain a lively action of the bowels. Give the Sweating powder every night, and plenty of Motherwort tea to drink. and bathe the whole body frequently with weak lye ; give the patient from three to five grains of Ipecac three or four times a day. If he has much pain of the bowels, you will simmer Catnip, Smartweed, Wormwood, Hops, &c., in vinegar, and apply to the parts ; warm and change frequently.

In the latter stages, when the patient is much reduced, and the fever abates, you will stimulate ; give wine, or wine and water.

Should the stomach assume a putrid state, you may give him two or three table-spoonsful of yeast, three or four times a day.

Regimen.—Avoid all stimulating drink and food at the commencement of this fever, but when weak and reduced, give more nourishing food, and Wine and raw Eggs, Oysters, &c.

Scarlet Fever.

Symptoms.—Commences with a chill, like other fevers, nausea and sometimes vomiting, thirst, headache, eyes red and frequently swollen, pulse high, breathing quick; subsequently the flesh begins to swell, and a pricking sensation is felt; red blotches usually appear about the neck and breast, then sweep over the entire body; usually in two or three days perspiration takes place, and the eruption disappears, and the scale or cuticle peels off.

Remedies.—Very little medicine is required in this disease. Only cleanse the secretions of the body, with a cool diet, and prevent the patient from taking cold.

But a second form sometimes appears, in which the throat and mouth are much inflamed, and soon succeeded by greyish sloughs, and give the parts a speckled appearance, and render the breath fetid. This is a gangrenous form, for the patient often dies in a few days, or if he recovers it will be very slowly, accompanied by many unfavorable symptoms, such as Dropsy, Ulcers of the throat, nose, &c.

Remedies.—In this as other fevers, cleanse the stomach and bowels, give the Sweating Powder once in four hours, with the free use of warm Lemon, or Catnip teas, until he sweats; bathe the body in warm Saleratus water freely and frequently, and give Saffron tea frequently. You may also give three or four Ipecac powders during the day. This will not only relax the bowels, but promote perspiration, and is also an expectorant. Bathe the throat if swollen and painful, with the Sassafras Liniment; gargle with Shoemoke bolls or berries, 1 oz. to 1 pint of water, steeped strong ; also with Yeast and water ; weak Lye is also good. Apply Mustard poultices to the feet. Should the patient be very restless, give the Sweating Powder at night, or two Pills as large as a pea, of Motherwort Extract. Should there be symptoms of inflammation of the brain, apply warm wet cloths, or cold applications such as Brandy or Rain Water and Vinegar ; Camphor is also good.

Regimen.—Give glutinous drinks such as Slippery Elm, Flaxseed, Comfrey, or Marshmallow Root teas.

Food mild and cooling ; room neither cold nor hot ; keep the air pure. You may also give Snakeroot tea occasionally.

Infantile Fever.

Symptoms.—The child is peevish ; lips dry ; hands hot ; breath short and quick ; head aches ; pulse quick, often from 100 to 140 in a minute ; lays stupid ; sleep disturbed ; food rejected ; sometimes bowels are re laxed, and sometimes costive ; evacuations slimy ; often delirious and sometimes speechless ; generally drowsy ; sometimes seems quite well, but peevish. This fever is mild at the beginning, slow in its progress, and uncertain in its results.

Remedies.—As in many other fevers, give an Emetic ; then cleanse the bowels with Senna and Manna two or three times a week. Promote perspiration ; bathe the entire body with warm water ; apply cloths wet in vinegar and water, to the head, and a mild mustard poultice to the feet ; give Elder Blow tea freely ; also the Ipecac and Sweating powders. When the fever abates and the child is weak, give a Tonic, such as Chamomile Flowers, Gentian, or Colombo teas. Diet as in all fevers.

Inflammatory Diseases.

These diseases are characterized by redness, heat, pain and swelling of a part, or the whole body, either acute or chronic, and may attack any or either organ or member of the body.

Inflammation of the Brain.

The patient is attacked with fever symptoms, such as redness of the face and eyes, retires from light and noise, headache, delirium, pressure of the head, throbbing of the temples, feet cold and head hot, &c.

Remedies.—Bring the heat from the head to the feet as soon as possible, by cold applications to the head and warm bathing of the feet, draughts, &c. Give a Cathartic once in four hours, until an action is procured, and then repeat two or three times a week the Mandrake compound, and promote perspiration as in fevers. But should the symptoms not abate, you will apply a Mustard Poultice between the shoulders, and foment the head by applying Hops simmered in vinegar, warm, and change frequently, and if restless give 1 to 3 Motherwort pills at night.

Regimen.—Food cool and mild. Drink mucilaginous teas.

Inflammation of the Ear.

Symptoms.—The pain is very acute ; ear inflamed ; more or less fever, and sometimes delirium, attended with throbbing. Suppuration takes place, and often continues for years.

Remedies.—If the pain is very acute, simmer bitter leaves in vinegar and water, and foment the parts ; repeat until the pain abates. If this fails, take ½ oz. of Sassafras Oil, 1 oz. Olive Oil, and 1 drachm Camphor; mix all, and drop it on wet on a little cotton, and put into the ear. Onion Juice, Laudanum, &c., is also good ; the Vapor Bath is also good, applied to the head ; Hickory wood Sap is also good, as well as for deafness. If the ear suppurates, inject with an ear syringe Olive Oil, Castile Soap Suds, decoction White Oak, &c.

Mumps.

The glands of the neck upon one or both

sides, become enlarged, hard and painful; difficulty often of both breathing and swallowing; the swelling often extends to the testicle, and becomes very dangerous as well as painful, increasing for three or four days. It is frequently attended with fever.

Remedies.—Promote perspiration; give Cathartic (gentle). If very painful, immerse raw cotton in the following : Castile Soap, scraped, 1 drachm; Sassafras Oil, ½ oz.; Olive Oil, 1 oz.; Camphor, 3 drachms, apply to the parts, and give the patient warm tea freely. If the testicle is swollen, take equal parts of good old Jamaica Rum, Tinct. Camphor, and Laudanum; mix and warm a little, and bathe frequently; or wet cotton and apply. You may also give an Emetic, and Sweating Powder, &c.

Quinsy.

This disease is characterized by redness of the tonsils, difficulty in swallowing and breathing, hoarseness and dryness of the throat, fullness of the tongue, difficulty in expectoration ; it is attended with fever, pulse full and hard, often from 100 to 150.

This disease occasioned the death of Washington.

Remedies.—Give an Emetic, Purgatives ; steam the head with bitter Herbs and Vinegar, inhaling the same freely, and repeat frequently until the symptoms abate. Bathe the affected parts freely with the Rum Liniment. The Sassafras Liniment is also good. To the latter add a little Hartshorne. Gargle the throat frequently with Shoemoke liquor, viz. : 1 oz. of the berry and pint of water ; you may add a little Sage, Alum, Borax, &c. A little yeast is also good with the above. Saltpetre is also good. Weak Lye is also valuable. If the throat is badly swollen, make a Poultice of Flour, Slippery Elm, and Buttermilk ; apply warm, and change frequently. Henbane simmered in spirits, and applied, is also very valuable.

Regimen.—Avoid all cold or stimulating drinks ; food mild, as in all other inflammatory symptoms.

Croup.

Symptoms.—A whistling noise; frequently hoarseness, when coughing, like the barking of a hoarse dog; great thirst and restlessness; difficult expectoration.

Remedies.—Carry off the mucus as quick as possible by means of an Emetic, and cleanse the bowels as soon as practicable; bathe the feet in warm water, or apply Mustard Poultices to the same; repeat the Emetic every day, or oftener if necessary; also keep the bowels open; bathe the throat and stomach with the Sassafras Liniment; you may also steam the parts, and inhale Hoarhound, Catnip, and Hops, simmered in Vinegar, and bind them about the neck warm, and repeat. You may also give a tea-spoonful of the following: Onion Juice sweetened with honey, ½ to 1 tea-spoonful every hour. Diet the same as Quinsy.

Hooping Cough.

Symptoms.—A tightness of breathing; thirst; quick pulse, and some symptoms of fever; hoarseness; cough; expectoration difficult. It assumes this character ten to twelve days, and then is attended with a peculiar kind of hooping at intervals, and is quite violent, and in its advanced stages the patient frequently chokes, and then vomits either from the stomach or lungs, or both, and often bleeds from the nose or ears. It sometimes continues for months and even for years.

This disease is not curable; the symptoms can only be mitigated by proper remedies. To lessen the unfavorable symptoms you will give emetics frequently, sufficient to vomit lightly, so as to keep the phlegm as clear from the stomach as possible. Give the patient Poppy tea freely and frequently. Pennyroyal, Hyssop, Spearmint, Comfrey, Slippery Elm, and Flaxseed teas are all good. Give Castor Oil frequently, to keep the bowels open. Olive Oil is also good. Im-

merse the feet in warm weak Lye every
night. You may also apply mild solutions
to the feet and breast frequently. You may
also apply a plaster between the shoulders,
made as follows : 3 parts Hemlock Gum to
1 part White Pine Gum.

Regimen.—The food should be light and
easy of digestion, and drinks warm.

Colds and Coughs.

Symptoms.—It is not necessary to describe
the symptoms of this complaint; they are
too well known to be mistaken. I would
only admonish you, and say that this is the
forerunner of many of the ills of life. And
as the Apostle admonishes us to flee from
the appearance of evil, so I would say flee
from these attacks as soon as possible. No
time should be lost, lest a viper more terrible
fastens upon you.

Remedies.—First take the Steam Bath ; an
Emetic next ; cleanse the bowels, and drink
bitter Herb teas, such as Smartweed, Bone-

set, Hoarhound or Pennoroyal, Spearmint, Chamomile, &c. You may also make a Syrup of Onion Juice and Honey, or Poppies. Should your cold continue you may take the Cough Drops.

Regimen.—Avoid everything of a heating and stimulating character both in food and drink. Corn Meal Gruel or Bean Porridge is the best food. Keep warm, dry, and free from cold atmosphere. Drink warm teas freely, and immerse the feet in warm weak Lye each night on going to bed.

Inflammation of the Lungs.

Symptoms.—The patient is attacked with violent pain in the side or chest ; difficulty of breathing ; cough ; dryness of the skin ; heat, anxiety, and thirst ; pulse hard, strong, and frequent.

Remedies.—Employ a counter-irritant at once ; the best one is a mild Mustard Poultice ; leave it on until the skin is reddened, but not blistered. Give the Ipecac Tincture to loosen the mucus ; this will promote per-spiration, expectoration, and action of the bowels. You may also give the Cayenne Powder ; this will both lessen cough and inflammation, and give quiet and ease. Drink Comfrey, Flaxseed, Slippery Elm, or Marshmallow tea freely. Buttermilk Soup is also valuable. Immerse the feet every night in warm weak Lye, from twenty to thirty minutes.

Regimen.—The best food the patient can eat, is Buttermilk or Indian Meal Gruel ; these are both food, drink, and medicine. Avoid everything of a stimulating nature both in food and drink.

Pleurisy.

Symptoms.—The patient is attacked with chills, fever, thirst, restlessness, and then a sudden violent pricking pain in the side, ex-tending to the shoulder blade and back and sometimes over part of the breast, with fre-quent cough, expectoration, &c. The mucus thrown off at first is small in quantity, and then often having streaks of blood, and as it progresses it becomes more gross and more impregnated with blood. Pulse very strong and tight, like the string of the viol.

Remedies.—Give to an adult 2 tea-spoonsful of Sudorific Drops in Catnip tea, and if not relieved in half an hour, repeat. Bathe the feet in tepid Water or Ley. Apply Mustard Poultices to the side and Draughts to the feet. Camphor, Whisky, and warm Water, applied by wetting cloths, are all good ; or you may take 1 pint Alcohol or brandy, and 1 oz. of Cayenne Pepper ; simmer a little ; wet flannel and apply, and change as often as cool ; or, you may boil bitter Herbs in Vinegar, and apply warm, and change frequently. Give Pleurisy Root Tea freely through the day, say ½ oz. of the root to 1 pint of boiling water ; you must first bruise the root. Should the inflammation not abate, you may give 15 to 20 drops of Tincture of Foxglove, in Pleurisy Root Tea, and give the Sweating Powder each night.

Regimen.—Food cool, slender, and diluted. Avoid all stimulants, whether solids or liquids.

Inflammation of the Stomach.

Symptoms.—Burning heat, pain, and swelling, frequently hiccough, vomiting, cold extremities; hard, quick, and tense pulse; pain increased by pressure; thirst and pain increased by drink; restlessness and weakness.

Treatment.—To ease the pain, give 20 drops of the Cholera Compound, once in fifteen to thirty minutes, until the pain is relieved. Drink Slippery Elm Tea freely; Flaxseed, Comfrey, and Marshmallow Root Teas are all good. Foment the stomach with bitter Herbs, or if very urgent, apply a Mustard plaster. Give a dose of Castor or Olive Oil frequently, sufficient to keep the bowels rather relaxed. Should the patient continue to vomit, give Spearmint Tea and bind the herb upon the stomach.

Do not give an Emetic in this case; if you

do, you may endanger life. Carefully avoid eating or drinking anything of a heating, acrimonious, or irritating nature. Avoid all Spirits, Wine, or Malt Liquors. Food, Corn Meal Gruel, Buttermilk Pap, simple Toast prepared in water, mild Mutton or Chicken Broth. Drink, neither cold nor hot; it should be chiefly Slippery Elm Tea.

Inflammation of the Womb.

This disease often occurs from injury, such as Childbirth, Difficult Labor, Puncturing, Pollution, or Cold, and Obstructed Menstruation, &c.

Symptoms.—Heat of the bowels, burning pain, urine high colored and scanty, sensitive or painful to the touch, and sometimes bloating; the pulse is hard and frequent, countenance dejected, depression of strength, heat of the whole body, thirst, and sometimes a nausea.

Remedies.—Fomentations should at once be employed, such as Camphor Compound,

Hops simmered in vinegar and applied in a bag warm, and changed frequently. Catnip Tea should be drank freely. The Sweating Powder should be given from two to three times a day. Cathartic, give a small dose of Compound Mandrake three or four times a week, and give the Compound Tincture frequently and freely.

Inflammation of the Bladder.

Symptoms.—Acute burning and pain at the lower part of the abdomen ; a frequent desire to urinate ; a partial and sometimes entire obstruction of the urine ; frequent desire to evacuate ; a frequent and hard pulse ; sometimes vomiting &c., and frequently blood and matter passes with the urine.

Remedies.—Employ counter-irritants ; a Mustard Poultice, bitter Herbs simmered in water or vinegar ; or the Camphor Compound, warm, and change frequently ; inject warm water with a small syringe ; give from a ⅓ to 1 tea-spoonful of the White Drops three times a day. Also give the Compound Tincture of Burdock. Wild Carrot, Wild Parsley Seed, and Rush teas are all good. Diet, chiefly Gruel, Buttermilk, &c. Drink, cooling and mucilaginous, such as Slippery Elm, Flaxseed, and Pumpkin Seed teas, which are all good.

Small Pox.

Symptoms.—This disease is divided into two distinct classes. It is known as Distinct and Confluent. In the Distinct Small Pox the disease begins with an inflammatory fever. Cold stage, great languor and drowsiness. A hot stage soon comes on, and then is characterized by great drowsiness, and frequently a numb sensation, and often when the patient closes his eyes, it appears to him as if the entire members of his body were enlarged. From three to five days after the attack, the eruption appears and spreads over the entire body. At this stage the fever entirely subsides. At first, numerous red spots appear, and then gradually rise into pimples, which first appear upon the face, and at the sixth day are filled like a bladder. The fluid, and afterwards a pit, form at the centre. Between the eruptions the skin assumes a scarlet hue, the eyes red, face swollen, until the eyes are often closed. At the eleventh day the swelling abates, the pustules are quite full ; on the top of each a dark spot appears and then breaks, and a portion of the matter oozes out, and forms a crust or scab over the surface, and some days after, the scab falls from the skin, leaving a darkish spot.

Confluent form.—In this species the foregoing symptoms are much aggravated. The pulse is increased ; more fever ; increased drowsiness, and frequently delirium, and sometimes vomiting, and with the young frequently epileptic fits, and often proves fatal before the eruption appears, and when they do appear, they are in clusters like the measles, and often run together, and the entire face and other parts assume a black or crusty appearance. The eruption does not properly fill, but assumes a watery hue, and evidently indicates putrefaction. It is always attended with fever, and the more fever the greater the danger.

Remedies.—If vomiting, give the patient Spearmint tea warm and freely, and if violent and difficult to check, bind the herb warm upon the breast; or you may give him Saleratus water to drink; when this is accomplished, give the Mandrake Compound, and repeat once in four hours until the bowels are thoroughly cleansed. Then promote perspiration; give warm tea of Catnip and Saffron, equal parts; frequently immerse the feet in warm weak Lye, and bathe the surface of the entire body once or twice a day. If fever increases at any time, see that the bowels and stomach are kept well cleansed, and give the Sweating Powder every night. Should there be much pain in the head, apply Mustard Poultices to the soles of the feet, and bathe the head with equal parts of Rain Water, Vinegar and Spirits.

Should the throat be sore, gargle it with

equal parts of Hyssop and Sage, sweetened with Honey, and three or four times a day give a wine-glassful of yeast. Should the patient manifest restlessness, give him Virginia Snakeroot weak tea to drink at night.

Should he be languid and weak, give him Buttermilk to drink, not too sour, or made in Pap. A little wine whey, &c.

Should the eruptions not fill, you will give him Ale and Molasses well warmed, to drink freely. Should this fail, give him warm milk punch freely.

Or after the eruptions fill, should they strike in, you will pursue the same treatment.

Secondary fever.—This is the most dangerous of all other stages. This generally occurs when the eruptions begin to turn dark. Many are carried off at this stage. To correct this at once, give a Cathartic, even though the bowels may be relaxed. By this method you will carry off the impurities of the body.

The above treatment to be pursued in either of the forms of the disease.

Regimen.—If the pox is well filled, there is no occasion for alarm. Give him Corn Meal Gruel, Buttermilk, Roasted Apples, Mush, or Hasty Pudding. Let his diet be cooling and yet nourishing. Avoid all stimulants except in such cases as you are directed to use them. Keep the room well cleansed; a fresh air, but of moderate temperature. If there should be much irritation of the eruptions, Poultice with Buttermilk and Flower of Slippery Elm.

Measles.

Symptoms.—This disease is characterized by uneasiness, chills, shivering, headache, sore throat, heaviness, and sometimes vomiting, &c., but more commonly with heaviness of the eyes, redness, tears, and inflammation, resistance of light, and frequent sneezing ; the heat increasing rapidly, a dry cough, violent pain in the groins, and frequently looseness ; tongue coated, thirst great, &c. · the eruptions appear from the third to the fifth day, and the spots dry up ; the skin peels oft and a new one comes on ; from the ninth to the eleventh day, no redness is seen upon the surface, but unless the secretions and excretions have been well cleansed, the cough will continue and fever increase, and bring on great distress and danger.

Remedies.—I know of no better remedies or course of treatment for Measles, than that I have laid down under the head of Small Pox. Except that of Pulmonary symptoms, where the mucus has accumulated, you will give the patient a mild Emetic, and repeat if necessary.

Regimen.—Diet the same as for Small Pox. While recovering, eat light food, and in small quantities, for some time.

Delirium Tremens.

Symptoms.—Vomiting, belching of wind, &c. This disease is gradual in its progress, and is a number of days before it arrives at its worst stage. It causes restlessness, wakefulness, and a constant disposition to walk to and fro ; delirium ; spirits agitated ; sudden frights, and often visions of ghosts, snakes, devils, &c. It is attended with fever and costiveness, and often terminates in fits of epilepsy, but with proper treatment the patient may recover.

Remedies.—You should give the patient brandy or gin, which frequently affords immediate relief. Should the blood rush to the head, which may be known by the redness of the eyes, palpitation of the heart, &c., immerse the feet and limbs at once in warm, weak Lye, and give Motherwort tea freely. You may also apply Mustard Poultices to the feet and back of the neck. Also give Cathartics and Emetics, and 2 to 3 Motherwort Pills the size of a pea, on going to bed.

Regimen.—Care should be taken that during the fit he does not do violence to himself. His food should be chiefly Corn Meal Gruel, and he should be kept quiet and easy.

Cholera Morbus.

Symptoms.—The attack is sudden and violent, causing nausea, vomiting, pain in the stomach, accompanied by griping and pain in the bowels and purging. The stools are at first thin and watery, and often accompanied with green bile. As the disease advances, these symptoms are more violent, and cause a spasmodic affection of the muscles of the bowels and extremities. The patient is drawn nearly into cramps by every paroxysm, and often screams at the top of his voice from the most excruciating pain. Thirst great, but when gratified, causes vomiting. The pulse becomes feeble, small, and intermitting; extremities cold; countenance pale, and a cold sweat breaks out over the entire surface. It is a violent and dangerous disease, and often proves fatal in a few hours.

Remedies.—When first attacked, give from 20 to 30 drops of the Cholera Compound, and Motherwort tea freely; repeat every thirty minutes. Should these remedies fail to check, which is seldom the case, you may give the Neutralizing Powder, viz.: to 1 large tea-spoonful add half a pint of boiling water, and loaf sugar to sweeten, and when nearly cold, add 2 table-spoonsful of brandy, and give two tablespoonsful every half hour. In violent cases you may add to each dose a few drops of laudanum, and continue until the symptoms abate. Attention must also be given to the bowels, to which you may apply a small bag of Oats, simmered in water, warm, or Hops simmered in vinegar, or bitter Herbs, steeped, and changed, and applied warm to the parts. If costive and full of flatulency, take one pint of sweet Milk, and a quarter of a pound of hog's lard; warm and inject every few minutes until he gets an action of the bowels. Immerse his feet in warm Ley or Water, and apply warm bricks or bottles to his extremities, and promote perspiration. Give him to drink all the Oat Meal or Indian Meal Gruel he can conveniently drink, a little warmed. After the patient recovers, in about twelve hours give him a mild Cathartic.

Asiatic Cholera.

Symptoms.—They are much the same, only in a more violent form, and consequently should be treated much the same as Cholera Morbus, only as the symptoms are much more urgent, of course the doses should be larger. The Cholera Compound was used during the prevalence of Cholera in Cincinnati, with almost perfect success, and my opinion is, that no remedy can be found in either of the above diseases, more efficient.

Cholera in Children.

This disease is known as Cholera Infantum, and resembles somewhat Cholera in adults. Although there are many variations, it is more commonly known as Summer or Bowel Complaint. It frequently attacks children during the summer, and is often occasioned by eating green corn and fruits, and by teething, &c.

Symptoms.—It commences with a mild fever; diarrhœa, nausea, and subsequently vomiting; stools very offensive, of a slimy, whitish, frothy, or colorless, watery fluid. In its progress the child begins to sink, becomes very weak and pale; extremities cold; skin dry and shriveled, and heat of the head and bowels; eyes sunk and dull, pulse weak and irregular; patient dull and sleepy, and when asleep, eyes partly open. This disease may assume a chronic form, and the child become a mere skeleton and linger a number of months; or it may die in a few days, unless the most efficient remedies are employed.

Remedies.—If the patient is vomiting, give the neutralizing powder. Dose according to age and urgency of the case; this, however, will be required in an advanced stage. If when first attacked you will give the Cholera Drops in proportion to age and strength, you will invariably check the disease in one hour; but you should always give a Cathartic after the disease is checked. If the skin is dry, bathe the entire surface with warm weak Lye or Saleratus Water,

once or twice a day, and give the Sweating
Powders every night, together with Catnip
tea, and drink freely Slippery Elm, Comfrey,
or Flaxseed teas.

Regimen.—Keep the atmosphere pure, and
clothes clean. The best diet is boiled milk,
thickened with wheat flour; it may be spiced
with a little nutmeg, cinnamon, or cloves.

Vomiting.

Symptoms.—Extreme loathing of food;
nausea; vomiting; general debility; lassi-
tude; frequent straining and cramping of
the stomach, returning about once every
hour; great distress and burning of the stomach, and frequently hiccough and belching
of wind; tongue slightly furred; breath
offensive, &c.

In a more advanced stage the patient becomes stupid, and sometimes the eyes set in
the head, and occasionally vomiting of black
matter, and often destroys the patient in a

few hours. He is dizzy, trembles, stag-
gers, &c.

Remedies.—To allay the irritability of the
stomach, give Spearmint tea, and add a little Saleratus, say 1 tea-spoonful of Saleratus
to 1 pint of the tea, and give a table-spoonful every hour until relieved ; when the stomach is relieved, give the Neutralizing Powder, until it produces a little action of the
bowels. Should this fail, you may give the
Mandrake Compound. Smartweed tea is
said to be good ; or you may give an Emetic
in the beginning ; or the Cholera Drops,
which have cured in a few minutes.

Water Brash.

Symptoms.—Usually attacks in the morning and forenoon, when the stomach is emp-
ty. It comes on with pain in the stomach,
similar to cramps, after which a great deal
of water is thrown from the stomach, of an
acid and sometimes insipid taste, and is
sometimes ropy, resembling the white of an

egg, and after awhile passes off. This disease seldom proves fatal.

Remedies.—Take 8 oz. of Compound Tincture of Senna, ½ an oz. of Balsam of Tolu Tincture ; mix, and take 1 table-spoonful every morning.

Regimen.—Avoid all greasy and acid foods. Pepper, Horseradish, Mustard, &c., may be eaten.

Cramp in the Stomach.

You will immediately apply a lively friction to the stomach, and take 20 drops of the Cholera Compound, and repeat in fifteen minutes if necessary. You may also take Catnip, Motherwort, or Smartweed, or Set Teas.

Heartburn.

When attacked, take 1 tea-spoonful of Carbonate of Magnesia, morning and evening, in milk. Or you may take a mild

Emetic, and avoid eating anything which produces flatulency.

Hiccough.

Causes.—Inflammation of the stomach; flatulency; drinking ardent spirits; swallowing tobacco juice. If from either of the two latter causes, give half a table-spoonful of Hog's Lard and one-tea-cupful of sweet Milk warmed. Or a mild Emetic. Should it be from inflammation, give a little Laudanum, after which give Slippery Elm tea freely.

Bleeding of the Nose.

This may generally be stopped in a few minutes by holding a piece of silver between the teeth, or applying cold water to the head and nape of the neck. But sometimes these remedies fail. You will then wet cotton in brandy, and saturate it in pulv. alum, and plug the nose. I was called in one day to see two ladies who were bleeding at the nose until they had bled nearly to death, and soon succeeded in stopping the blood with the above remedies. Grated dried beef is said to be good ; plug the nose with it. It may be important in some cases to immerse the feet in warm lye.

Jaundice.

Symptoms.—Loss of appetite ; dullness ; costiveness ; soon the skin turns yellowish, first of the eye, then the nails ; urine high colored ; stools grey ; skin dry, with frequent pricking sensations ; generally drowsy and sleepy ; sometimes over wakeful ; as the disease advances, the skin becomes more yellowish.

Remedies.—As this is an obstruction of the bile in its passage into the duodenum, you will first give an Emetic rather mild, and then in a day or two give a half tea-spoonful of equal parts of pulv. Mandrake Root, pulv. Cloves, and Cream Tartar, and repeat if necessary. Then take the following, viz. : 1 drachm Yellow Dóck Root, 2 drachms Bitter Root, 2 drachms White Poplar Bark, 1 drachm Capsicum, 1 drachm Wild Cherry Bark ; cover with boiling water ; then add 1 pint Holland Gin. Dose, $\frac{1}{2}$ to 1 wineglassful three times a day. At the same time take No. 4 Powder each morning, from 3 to 5 grains, or sufficient to keep the bowels a little relaxed.

After the disease is thoroughly broken up, you may take the following : 2 drachms each of Spikenard, Comfrey, and Sol Seal, and 1 drachm of Cherry Bark, 1 drachm of Babary, one of Columbo, and 1 of Chamomile Blows ; add boiling water ; let stand one hour ; then add 1 quart of good old Port or domestic Wine, and take from a half to one wine-glassful three times a day.

Inflammation or Ague of the Breast.

Symptoms.—Redness, Swelling, Pain, Obstruction of Milk, &c.

Remedies.—Drink Yeast and Water freely until it purges; at the same time apply the Rum Liniment, to which you may add a little pulv. Black Pepper. Wet cloths and apply warm and keep moist. But should you fail to remove the inflammation with this remedy, then there has evidently matter formed, and the treatment must be changed. You will then poultice as follows: take Slippery Elm and weak Lye, make a poultice, and change frequently. You may apply a little Oil or Butter to the parts first, to prevent from adhering. Should the pain be intolerable, you may take now and then a pill or two of Motherwort Extract as large as a pea, or an Opium pill the same size. You may also give the Sweating Powder at night, and immerse the feet in warm water. Promote perspiration, and if necessary have the breast opened with a lancet. It is better, however, to be patient and let Nature do its own work. After the breast has suppurated and the inflammation has subsided, you may apply any healing salve until well. Diet, the same as in other inflammatory cases.

Locked-Jaw.

Causes.—This disease proceeds from violent cold, stabs, punctures, bruises, cuts, &c., which must in every case demand our first attention.

Treatment.—Poultice with Slippery Elm and weak Lye, or steam with bitter Herbs the affected parts if much inflamed, and repeat three or four times a day. Or you may immerse the parts in weak Lye. Also take the Vapor Bath, immerse the head as well as the body ; set until you sweat freely, say from fifteen to twenty minutes. At the same time give an Emetic. You may add a little Capsicum. Repeat the dose every twenty minutes until he vomits freely. If the bowels are constipated, give an injection of hog's lard and sweet milk, and after the stomach is thoroughly settled, you may give 1 oz. of Olive or Sweet, or Castor Oil, and repeat if necessary. You may also give the Sweating Powders, two or three a day, and frequently drink warm Catnip, Hoarhound, Penneroyal, or Motherwort tea.

Inflammation of the Eye.

First remove the cause. It sometimes happens that dust, or sand, or cinder, or an eyelash gets into the eye, and produces inflammation. Should this be the case, you will turn the lid outward, and remove the cause. But from other causes, such as scrofula, injuries, heat, bright light, bleak winds, the use of liquors, or from measles, small pox, or syphilis, then you will poultice the eye at night with pulverized Slippery Elm and Buttermilk. During the day anoint with the Brown Ointment three times a day. Or you may use an Eye Water made as follows: Take the pith of Sassafras Root, extract its strength in rain water, add equal parts of Spirits Camphor and Laudanum. Wash the eyes frequently. Give an Emetic every week and Cathartic twice a week. The Vapor Bath is also good. You may also give the Sarsaparilla Syrup.

Felon.

The first thing to be done is to lessen inflammation and pain, when this has taken place. In its first stage, however, it may frequently be removed by immersing the finger in warm weak Lye, and holding it there a long time. But should it continue obstinate, you will take Tansy, Catnip, Hops, Smartweed and Wormwood; boil fifteen to twenty minutes; add half a teacupful of Soft Soap. Place the hand over it. Cover so as to confine the steam until the hand sweats freely, and repeat three or four times a day, and at intervals you may poultice with Flower of Slippery Elm and weak Lye. Poultice until you discover a small white spot; then take a needle and press gradually until it enters the matter. Should fungus flesh follow the matter, you will apply a little of the Caustic or Potash, and you may introduce a little Caustic daily. It then may be dressed with a little Drawing Salve of any kind. You may now and then take a Cathartic, and let your diet be light and cooling.

Chilblains.

These are occasioned by the parts being frosted, and may be corrected at once when it happens, by immersing the parts in snow or cold water, after which, let brisk friction be applied. Then take 2 oz. of Olive Oil, 1 drachm of Hartshorne, and 1 drachm of Camphor Gum. Mix and shake well. Bathe the parts two or three times a day, and immerse the feet in Lye at night. You should wear linen stockings winter and summer.

Boils.

When highly inflamed and painful, you may steam with the bitter Herbs, and make a poultice of Slippery Elm and Flaxseed Oil; first boil the Elm in rain water or weak lye; after it suppurates and breaks, apply a little healing Salve. To prevent Boils, you may make a syrup of equal parts Yellow Dock and Burdock Roots, and take from a half to one wine-glassful three times a day.

Ague in the Face or Jaw.

This complaint frequently comes from colds and other causes. The face becomes swollen and painful, and the pain often becomes intolerable.

Remedies.—Bathe the parts with the following preparation, viz.: Equal parts of Hartshorne, Sulphuric Ether and Alcohol. Or you may steam with bitter Herbs, and you may immerse a bit of cotton in the Tinct. of Capsicum, and place between the cheek and teeth.

Inverted Toe Nail.

Immerse your feet in warm, weak Lye; then poultice with Slippery Elm until the inflammation is reduced ; then cut the nail at the side affected a little from the flesh, and carefully skin it from the parts until you can

withdraw it by means of a pair of tweezers.
Should there be proud flesh, apply a little
Caustic and Healing Salve.

Corns and Warts.

How to Cure.—Take ¼ lb. of Potash, 1
drachm Extract Belladonna, and 2 drachms
Gum Arabic; dissolve in a little water;
add wheat flour, and work all into a paste;
apply a little to the corn, then work around
it with a sharp penknife, and soon it will
come out; then apply a little Olive Oil or
Vinegar, and be careful not to take cold after it for a few days.

Scalds and Burns.

Remedies.—Immediately when the accident
occurs, you will pulverize charcoal and mix
with hog's Lard, and spread on a piece of
linen and apply to the parts. This will relieve the pain in a few minutes. But should

you fail getting this remedy in time, until the parts become inflamed, you will then poultice with Flour of Slippery Elm. First apply a little Lard, fresh Butter, or Olive Oil, so as to prevent the plaster from adhering to the parts, and thus continue until well.

Ringworm.

Symptoms.—This disease begins with a small red spot, sometimes not larger than a three or five cent piece, and continues to spread until it becomes as large, frequently, as the palm of the hand. When the blood is heated by exercise, the parts itch intolerably, and the sore is greatly aggravated by scratching, so that the patient seldom enjoys rest or comfort.

Remedies.—Take Fire Weed Oil or Herb bruised, and mix with hog's Lard; after washing the parts with Castile Soap and water, bathe the parts two or three times a day. Or you may take pulv. Yellow Dock

Root, and mix with Lard, and apply in the same form.

Itch.

Causes.—It is generally taken by coming in contact with some person having it, or wearing their clothes, or sleeping in the same bed, &c. But it is frequently the result of impure air, improper food, and filthy, dirty clothes, houses, &c.

Symptoms.—It first attacks the joints between the fingers, and then continues to spread until it sweeps frequently over the entire body. The pimples are first small, but frequently increase in size until they become like small boils. It is said that they contain animalculæ, which may be seen by means of the microscope.

Remedies.—This disease may be cured as follows: Make a syrup of equal parts of Yellow Dock and Burdock Roots, and take from a half to one wine-glassful three times a day, and anoint the parts two or three

times a day with the same, as is laid down
under the head of Ringworm.

The old fashioned way of curing Itch is
as follows: Adult—take 1 tea-spoonful
Cream Tartar, and 1 tea-spoonful Sulphur,
mixed with molasses, three mornings, and at
the third night anoint the entire surface of
the body, and warm in well. Wrap in an
old blanket, and sleep soundly until the next
morning. Rise and wash in warm water
and Castile Soap Suds. Avoid taking cold
for a day or two. I have never known this
to fail when properly applied.

Although I have known many bad results
from the use of these as well as all other
minerals, I can only say, if you employ them,
be careful that you use them judiciously, and
not take cold afterwards.

Sprains.

If much swollen and inflamed, you will
take Smartweed, Wormwood, Mayweed, and
Hops; boil well; place the parts over the

vessel ; cover well so as to confine the steam until it sweats freely, and repeat as often as painful, and bathe frequently and freely with the Sassafras Liniment. If after the pain abates you find the joint weak, bathe the parts with the following : Take the inner bark of White Oak, boil a strong liquor, and bathe three or four times a day.

Antidote for Poisons.

There is not a family on earth who should not understand the nature and cure of Poisons. They are frequently taken into the stomach accidentally, and their effects are often so sudden and violent, that there is no time to procure a physician, and often when called, he does not know how to cure, or in his practice has not his remedies. It is a consoling thought that no great skill nor foreign remedies are requisite to remove poisonous substances from the stomach.

Nature is the best doctor in the world, if we will only be guided by its dictates.

Hence when poison is taken into the stomach, it soon occasions sickness and an inclination to vomit. Thus we are admonished at once what must be done, namely, cleanse the stomach by means of an emetic.

There are, however, several classes of Poisons, which I shall notice. They consist of mineral, animal, or vegetable mineral, and are of an acrid or corrosive nature, as Arsenic, Zinc, Iodine, Corrosive Sublimate of Mercury, Antimony, &c.

Remedies.—Immediately after you discover the accident, you should resort to one of the following remedies : Drink large quantities of warm Water and hog's Lard, or sweet Milk and Salad Oil ; or fresh Butter may be melted and mixed in the milk ; or you may give 1 tea-spoonful of pulv. Tobacco, mixed in molasses, and repeat every ten minutes, and drink large quantities of warm water. Either of the above remedies must be urged and repeated until the patient vomits freely. Or you may give a Lobelia Emetic, or the Emetic Compound. But in every case double the usual quantity must be

administered.   Pulv. Mustard in water,
given warm, will also cause the patient to
vomit it from the stomach, and in each case
be sure to give the Oil, Fat, or Butter, with
the other preparations.

I mention all these ingredients, in order
that you may find some of the above in that
moment of peril and fright.

Vegetable Poisons are generally of a nar-
cotic or stupefying nature, as Poppy, Hemlock, Henbane, Berries of Deadly Night-
shade, Opium, &c.

Remedies.—The same as for minerals are
to be employed, and after the poisonous substances are removed from the stomach, it
would be well to give gentle purgatives,
such as Olive or Sweet Oil, and should the
patient be weak, Tonics or Wine to drink,
and let his food be of a mild quality, Flaxseed, Comfrey, or Slippery Elm teas freely
for two or three days. It would be well,
however, to note in addition to the above,
that sometimes by taking Opium or Laudanum, the patient becomes stupid and difficult to arouse.   In that case it is important

to use other means to arouse and keep him awake. In this case he should be shook, tossed, and moved about. You may apply pure Mustard Plasters between his shoulders, legs, or arms, and apply stimulants to his nose; Spirits or Salts of Hartshorne may be held to his nose, or you may let him snuff Cayenne Pepper, &c. You may also give him strong Coffee to drink freely, or you may give him a large tea-spoonful of pulv. black Mustard, mixed with water, and repeat if necessary in ten to fifteen minutes.

Ivy Poison.

Both man and beast are liable to get poisoned by a vine very common in this country, known as Poison Ivy, Poison Weed, or Mercury. It produces great heat, itching and burning, swelling and inflammation.

Remedies.—Take the Bark of Sweet Elder and simmer in Buttermilk; wash the parts freely and frequently, after which you may apply Sweet Oil. If very irritable, poultice

with Slippery Elm. You may also use the
Fire Oil Ointment three or four times a day.

Bite of Snake.

Remedies.—Take the roots and branches
of the Red Plantain and Hoarhound, equal
parts ; bruise all in a mortar ; then squeeze
out the juice and give as soon as possible a
large table-spoonful of the juice, and if necessary, repeat in one hour, and apply a tobacco leaf to the wound ; change frequently.
Or you may apply the caustic to the wound
and then the tobacco leaf.   Salt is also
good.   The patient should also have frequent
Cathartics.

The following is good :   Take a bottle of
Spirits of Turpentine, place the mouth over
the wound, until the pain is extracted, which
will be in a short time.

Bite of a Mad Dog.

Remedies.—Take ash colored, ground, or pulv. Liverwort, ½ an oz., Black Pepper ¼ of an oz. Mix all and divide into four parts. Take one each morning in a half pint of sweet Milk, warm, and after the above medicine is taken, then the patient must be dipped under water head and all, and not remain in over half a minute, for thirty days, and then for two weeks three times a week. This is old Dr. Mead's remedy. In this I have but little confidence.

Dr. Buchan recommends if there be no blood-vessel injured, the parts adjacent to the wound may be cut out. But this must be done soon. But if not practicable, mix Salt and Vinegar, and apply, after which, take Yellow Basilicon mixed with Red Precipitate, and apply twice a day.

Dr. Beach recommends the following, and I have more confidence in him than either or all those noted. First have the wound cupped as soon as possible, after which apply the Caustic Potash until an eschar is formed.

Then apply an Yeast Poultice, and keep up a discharge as long as possible. Then take a strong infusion of Scullcap through the day, and Mandrake Compound once a week. But if there be symptoms of Hydrophobia, take a Lobelia Emetic every other day, and at the same time take the Vapor Bath.

Myself together with two young gentlemen, were bitten by a dog having Hydrophobia, among the Dutch near Schenectady, N. Y., one in the nose, one in the left hand, and the other in the right. A Dutchman orescribed the following strange prescription: Equal parts of the false tongue of a colt, and the jaw bone of a dog, and verdigris, to be taken nine mornings in succession. Also from one quarter to one third of a copper, to be filed, and taken during the day. We all took the above, and neither myself nor they, so far as my knowledge extends of them, have had the disease. I believe, however, they had the wounded parts cut out, but I did not.

I am of an opinion that the Red Plantain would cure or prevent the disease, by eating large quantities of the stalk, of the leaf, or squeezing out the juice from the stalk and root, as prescribed for the bite of a rattle-snake, but in that case it would be well to apply Dr. Beach's remedies externally.

Diseases of Children.

Infants are frequently afflicted with flatulency and gripes, more frequently from the mother eating improper food than from any other cause. She should therefore be careful about her diet. When the child is thus attacked, you may give it some Catnip, Peppermint, or Penneroyal tea. But the best remedy in all cases of griping, teething, nervousness, and disquietness, is to make Syrup of Motherwort, and sweeten, say about the consistency of laudanum, and give the child from ten to fifteen drops, and increase if necessary. This will also cure fits in children. With this remedy I cured one in St. Lawrence County, N. Y., who had been afflicted with them two years, after all others had decided the case incurable.

PURGING OR LOOSENESS.—This may soon be corrected by giving the child from one to three drops of the Cholera Compound, or by giving both the mother and child the Neutralizing Powder.

TEETHING.—This complaint has cost the life of many a child. It causes heat and pain in the head, restlessness, fever, swollen gums, dysentery, and often fits.

Remedies.—It may be necessary to cut the gums a little, and administer a little Casto⸱ Oil every other day, unless the bowels are relaxed. You may also give the Mother-wort Extract, and Catnip Tea, and promote perspiration. You may give the child a crust of bread, or make a hole through a silver dollar, and hang about the neck.

SORES ABOUT THE EARS, GROINS, &c.— Wash the parts with Castile Soap and Water; wipe dry with linen; then you may sprinkle on the sores a little Flour of Slippery Elm, or apply the Fire Oil Ointment, made as follows : Take 2 drachms Oil and one quarter of a pound of pure hog's Lard. Mix cold.

SORE MOUTH.—Children are frequently afflicted with this complaint. Little small spots appear in the mouth. Give a gentle physic, and wash the mouth frequently with a tea made of Sage and Hysop, sweetened with honey.

CONVULSIONS FROM TEETHING.—You will at once immerse the feet in warm, weak Lye, and give the Motherwort Syrup. Onion or garlic may be bruised and applied to the stomach. If there is much heat of the head, you may apply a cloth wet with rain water, spirits and vinegar.

SORENESS ABOUT THE NAVEL.—Apply the same remedies as prescribed for Sores about the Ears, &c.

RUPTURE.—Lay the child upon his back; then press the tumor or protruded parts back, and make a plaster of the Extract of White Oak Bark, and apply; then a compression over it with a bandage, to keep it in its proper place.

Tongue Tied.

It so happens that sometimes infants cannot nurse, from the traenum of the tongue being contracted. In that case, and in that only, should there be a very small incision made with a lancet or a pair of scissors. This cut, however, must be very small, lest a blood-vessel be severed.

Parents are apt to think their child is tongue tied sometimes, when such is not the case.

For Sty of the Eye.

Take 1 tea-spoonful of tea in a small bag, turn a few drops of boiling water on the same, and apply at night, and if necessary repeat the next night. This will certainly cure, unless it be from scrofulous taint

----

For Worms.

Dry Egg Shell pulverized, and mixed in molasses. Give the child one tea-spoonful three mornings in succession ; then a dose of physic. This remedy will destroy the worms.

To Preserve the Eye, and Restore Partial Blindness.

Occasionally press the eye-ball by means of placing the thumb and finger next to the nose and the temple. Take 1 gallon of Water, 2 drachms of Cream Tartar, and 2 oz. of refined Sugar. Wash the eyes three times a day.

This simple remedy has restored the eyesight of those partially blind for many years.

Another Remedy for Felon.

Take 1 pint Soft Soap, and add slacked Lime until formed into putty. Fill a leather thimble, made somewhat larger than the finger, and insert the finger.

A LECTURE

ON THE

LAWS OF THE MANY NERVOUS DISEASES TO WHICH WE ARE SUBJECT.

In speaking of the Nervous System, anat-omists include those organs which are com-posed of a nervous or pulpy tissue. The nervous system in man is composed of two parts, that which is called the cerebro-spinal axis, which is the brain and spinal marrow, and thirty-nine or forty-two pair of cords called Nerves, which pass off laterally from the cerebro-spinal axis, and ramify over ev-ery part of the body. Secondly, the gan-glions and plexuses, with their various cords, branches, and filaments. Under the term encephalon, are included the contents of the cranium, which are the cerebrum or the brain proper, the cerebellum or little brain, and the medulla oblongata. These different parts are included under the name brain. The brain proper, or Cerebrum, occupies the upper part of the head; the cerebellum is next; below it posteriorly is the Medulla Oblongata lower still.

I would here remark that I cannot go into a description of the brain phrenologically, but I am fully impressed with the value of phrenology as a science, and would earnestly recommend to my readers, especially those who are skeptical as to its truth, the admirable works of George Combe and the Messrs. Fowler. " Combe's Constitution of Man," is a work that is above praise. His other works are exceedingly valuable. The writings of O. S. Fowler contain physiological and phrenological truths well adapted to the wants of our age, and eminently calculated to bless humanity. J. N. Fowler is said by good judges to be the best practical phrenologist in America.

French anatomists reckon forty-two pairs of nerves. Of these, twelve pairs draw their origin from, or are connected with the encephalon, and thirty come from the spinal marrow. Each of the spinal nerves consists of filaments destined for two distinct uses—motion and sensibility. They have two roots, one arising from the posterior, the other from the anterior part of the spinal marrow. Sir Charles Bell says that the anterior part gives rise to Nerves of motion, the posterior to Nerves of sensibility.

The series of ganglions and plexuses, with the nervous cords, fibres and filaments which unite them, are collectively termed the great Sympathetic Nerve. It is connected with each of the spinal nerves, and with several of the encephalic, but does not arise from either. The Sympathetic is considered the great system of involuntary nerves. The nerves of the brain and spinal marrow, with their various ramifications, are called the Nerves of animal life. These are distributed principally to the muscles of voluntary motion, and to the sensitive surface of the body or external skin. The Sympathetic or Ganglionic Nerves are called Nerves of organic life. The ganglions of the Sympathetic Nerve give off branches, some ol which connect the ganglions with each other, and some interweave and inosculate, and form plexuses. From these, numerous branch- es are given off to supply the different or- gans with nerves. Besides the more deeply seated ganglions connected with the princi- pal viscera, there are two series of them which range along the anterior side of the spine, connected by nervous cords which ex- tend from the lower extremity of the spine to the base of the cranium, and enter by small branches through the carotid canal, along with the artery, and form connections with the fifth and sixth pairs of the nerves of the brain. These two series of what are termed peripheral ganglions, with their con- necting cords, are called Sympathetic Nerves, because they are believed to form the most intimate union of sympathy between all the viscera concerned in organic life.

At the base of the diaphragm, on the an- terior side of the spine, are two large gan- glions called Semilunar Ganglions. These give off numerous large branches, which, to-

gether with several from other parts, and
some from within the cranium, form a very
large central plexus in front of the spine,
which constitutes a kind of common centre
of action and sympathy to the whole system of organic nerves. This is called the
Solar Plexus. From this, branches are
given off in every direction, uniting with
nerves from the brain, and supplying the
different organs, particularly the stomach
and arteries. These are invested with a
lace-work of nerves, which accompanies
them to their termination in the glands,
skin, and mucous membrane, and other mem-
branes.

The Cerebro-Spinal Nerves are instruments of sensation and perception.

The Sympathetic or Ganglionic Nerves
are instruments of sympathy, and in a healthy state are not instruments of sensation ;
but in a diseased state they have great morbid sensibility, and a morbid sympathy may
also be induced.

The Nerves of the Bones in a state of
health, convey no appreciable sensation to

the brain. But bones may become diseased, and no pain is more acute than the pain of diseased bones. They may result in abuses of the nerves, and render them acutely sensible. The Nerves of Sensibility partake of the injury. Thus there is disease from abuse and disease from sympathy.

A great physiologist, from whose works these views of the Nervous system are taken, has said that the proper performance of the functions of life, and the welfare of each and every part of the system, depend upon the integrity of the nerves in supplying the necessary vital energy, and this again depends on their healthy state. By inducing a diseased condition and inflammation of any part, a new and abnormal centre of action may be established, equal in the power and extent of its influence, to the importance or the part, and the degree of its morbid irritation, which will not only derange the functions of the part itself, but also to a greater or less extent those of the other parts, and sometimes of the whole system, causing an undue determination of the fluids to itself, and resulting in morbid secretion, imperfect assimilation, chronic inflammation. disorganization by change of structure, by softening or induration, producing scirrhus, ossification, calculi, ulcers, cancers, and dissolution, or mounting into a high state of acute inflammation, and in a more violent and rapid career, bringing on gangrene, general convulsions, collapse, and death.

After carefully studying these views of the Nervous System, you will be better able to understand how we are affected by hurtful influences. In the first place it will be proper to lay down a definition of tone, which is that state of the nervous system when it responds with sufficient promptitude, vigor, and regularity, to the healthful and natural stimuli. Want of tone is of two kinds; first, when from deficient excitability the nerves do not respond with sufficient promptitude, vigor, and regularity, to the natural excitants, and the functions of the system, in all or in part, fall into a state of torpor. The second species of deficient tone, is when the nerves from excess of excitability, re-

spond too promptly and often irregularly to
the ordinary stimuli, and often act with violence, from the impression of causes which
in their normal condition affect them but
little if at all. It is this latter species
of deficient tone with which we have principally to do. It is produced by over excitement, mental as well as physical ; by
over exertion of the organs, without sufficient intervals of rest ; by whatever reduces
the physical energies of the system ; deficient
exercise ; deficient food ; mental and moral
indolence, as well as by excessive mental labor ; excessive evacuations ; and by whatever impairs or vitiates the nutritive functions of the system, as excessive, improper
or deficient food, improper drinks, vitiated
and confined air, deficiency of sleep, the de-
pressive passions, &c. In regard to the ex-
tension or diffusion of morbid action, this
takes place through the nervous centres ; irritation of the stomach by being reflected
upon the heart and lungs, hurries the respi-
ration and circulation ; irritation of the ute-
rus, by being reflected upon the stomach,

causes sickness, gastrodynia, &c. ; or upon the spinal nerves of motion, hysteria and neuralgia ; when upon the nerves of sensation, a piece of indigestible food in the stomach of a child, gives rise by reflection upon the nerves of motion or of animal life, to convulsions ; a portion of a briar in the end of the finger, by a similar reflection, causes tetanus, &c.

The world has so long looked upon passions misdirected, or excessive inaction, that many seem to have come to the conclusion that certain passions or propensities are inherently bad, and that they should consequently be eradicated. Now if we look in to this subject, we shall find that it is only the excessive or erratic action of the passions, that is productive of evil. These remarks are especially true of that appetite, instinct, or passion, which impels us to the propagation of our species. When kept within bounds, and exercised according to the dictates of nature, of reason, and of virtue, it has not only a beneficial influence upon the health and longevity of the system, is not merely to promote our individual happiness, and fulfill an important law of our being—increase and multiply, but it has a tendency to soften and improve the heart, and by the new relations thus resulting, to promote feelings of kindness and benevolence, and to interest us more deeply in the happiness and well-being of our fellow creatures. But the instinct of which we are speaking, is one which requires to be watched with the greatest care ; its tendency in the present artificial state of society, is to premature, excessive and destructive indulgence, and to this cause are to be attributed very many if not all of the usual diseases, which, instead of being confined as formerly to those classes which revel in luxury, commence now to inflict their pains and penalties upon the sex at large.

No form of nervous excitement is as injurious as solitary vice. The reports of our hospitals for the insane, if we had no other means of obtaining information, would convince us that this vice is exceedingly com mon. I shall proceed to show some of its effects, and then point out its causes, and the means of preventing it. That the unnatural, precocious, or excessive development of the sexual instinct, is disease as much as fever, and should be treated as such, I am fully persuaded. If hospitals were built for the social and solitary licentious, instead of casting them out from society, and suffering them to herd in dens of infamy, destroying and destroyed society might be in a more healthy state; but such is the excessive and diseased development of the animal nature of man, that the civilized world might well be turned into an hospital for the cure of diseases caused by licentiousness.

In the reports of our lunatic hospitals, masturbation or solitary vice ranks next to alcohol in producing insanity. All the diseases caused by social licentiousness, are produced by this form of nervous abuse.

I would remark that many of these diseases may be produced by other causes. I have given advice in almost every form of disease induced by this vice. I have seen idiotcy and insanity caused by it, and I think it is time that something should be done to rescue the most moral and conscientious, and sometimes the most promising youth from the mind-wasting ravages of an indulgence, of the terrible consequences of which they have never been forewarned; that is, it is the vice of ignorance not of depravity. The sufferers are personally less offenders than victims. That is a truth to be remembered. We should labor in the spirit of love not of blame, for the restoration of fallen, diseased humanity. Children are born with the impress of sensuality upon their whole being, in consequence of their parents. They are trained in a manner destructive to health, and it would be indeed a miracle if they should escape this vice. I am unwilling to leave this subject without again calling attention to the diseases which are caused by this habit. There is hardly an end to them. Dyspepsia, spinal disease, headache, epilepsy, and various kinds of fits, which differ in their character according to the degree of abuse and consequent disease of the nervous system. Impaired eyesight, palpitation of the heart, pain in the side, and bleeding at the lungs, spasm of the heart and lungs, and sometimes sudden death, are caused by indulgence in this vice. Diabetes or incontinence of urine, fluor albus or whites, and inflammation of the urinary organs, are induced by indulgence in this practice. Indeed this habit so diseases the nervous system, and through that the stomach and the whole body, that every form of disease may be produced by it through these disorders, and may afflict those who slightly indulged in the habit. Some who have been in a degree enlightened on these subjects, have feared to have others enlightened, lest it should increase the evil. They say there is safety in ignorance. I answer, the silent course has been pursued, till our world has become one vast pit of corruption. Has the world been safe in its ignorance? If not, will it be so hereafter?

Deslandes says that St. Vitus' dance is also at times caused by this vice, and I believe it. Deslandes and Tissot contain abundant evidence that the worst forms of spinal disease are occasioned by matsurbation. But light has dawned upon us, and we should be thankful for the blessing.

I have it from good authority, that among the insane admitted into the lunatic hospital from this cause, the proportion of females is nearly as large, and from my own observation, larger than that of males.

The reports of our lunatic asylums furnish melancholy evidence of the prevalence and increase of this vice. In the Fifth Annual Report of the State Lunatic Hospital at Worcester, Mass., we find the following :

The number of insane from masturbation (self pollution), has been even greater than usual the past year, and our ill success in its treatment the same. No good whatever arises in such cases from remedial treatment, unless such an impression can be made upon the mind and moral feelings of the individual, as to induce him to abandon the habit. In this attempt, even with the rational mind, we have to encounter mistaken views as well as active propensities. No effectual means can be adopted to prevent the devastation of mind and body, and the debasement of moral principle from this cause. Till the whole subject is well understood and properly appreciated by parents and instructors, as well as by the young themselves, how many of earth's noblest, even the brightest and best of our youth, have sunk beneath slow, wasting, nervous disease, the cause of which was neither known nor suspected by themselves or their friends. They have felt that they were doomed, that a destiny from which they could not escape, held them in its relentless grasp. They have shrunk from the struggle of life as if they were all nerves, and as if each nerve was bared to the pitiless pelting of the storm of life. They have felt sure that they were born with a constitutional nervous sensibility ; that life is a burden and a curse ; and often they have sought refuge in voluntary death, as a relief from sufferings that it was not in humanity to bear through. There are many causes for nervous diseases, still we have good reason to believe that many who rise every morning, like an infernal frog out cf Acheron, covered with the ooze and mud of melancholy, may trace their misery to this cause. Is he the true philanthropist, nay more, is he a christian, who, knowing all this, can be silent, can put his finger on his lips and say this subject is too delicate to be meddled with—you will but increase the evil by your efforts. Let ministers, let christians cease to denounce theft and murder; let them blot from the blessed book the commands against licentiousness, and use an expurgated edition of the Bible, lest the reading of the Holy Scriptures increase the evil.

When I reflect upon the distressing details to which I have listened, and the dreadful consequences I have witnessed, such as par-tial blindness, deafness, coughs, raising or vomiting blood, scrofula, eruptions of the face, breast and back, stiff joints, enlarged necks, glimmering of the eyes, cold extrem-ities, catarrh, insanity, heart disease, pale-ness, weakness, female weakness, fits, emacia-tion and death, from this solitary vice, I feel as if no time should be lost by physicians, parents, and even ministers from the sacred desk, alike should engage in this work of reform. I have been traveling nearly two years, treating chronic diseases, and it is universally admitted that I can and do describe the symptoms, and locate the disease of each patient as correctly as they could themselves, as well as to detect the cause or origin. And here I would unequivocally declare, that according to the best of my knowledge and belief, eight out of ten who die of consumption from the age of 16 to 28, started at this very point, and that two-thirds of all the chronic diseases from 16 to 30, is the legitimate cause of this practice. And if this be true, what a heart-rending picture, and what a terrible harvest to meet in the judgment. What a scene of horror, when we reflect that he that soweth to flesh shall of the flesh reap corruption.

My intention in this Lecture is to show the best means of preserving health. Health is a precious boon ; without it you may possess all other good, yet you are miserable. Were men half as solicitous about their health as they are about other interests, the amount of human suffering would be greatly mitigated. Luxury, intemperance, extrava gance, and improper conduct, are the seeds sown of a dreadful harvest, and invariably prove to be the common destroyers of our race. It is natural for us to indulge in our propensities and appetites, until some derangement of our digestive functions is the consequence, and then, instead of correcting our habits, we resort to poisons and nos-trums, which almost invariably create the disease for which they are taken to cure.

One ounce of prevention is worth a pound of cure. How many have you heard say they would live well, eat and drink what they desired, whether their life were short or long. A certain man within my own knowledge, of a dyspeptic habit, would eat a full meal, then run his finger down his throat, and vomit it up. And this is not an isolated case by any means ; they abound everywhere. Says Dr. Mott: All who have abused their stomachs will be brought to an account sooner or later. And adds : I am not sure but more disease results from intemperance of eating than drinking.

Unwholesome food and irregularities in diet, occasion many diseases. The whole constitution may be changed by diet. The fluids may be attenuated or condensed, rendered mild or acrimonious, coagulated or diluted to almost any degree.

Nor can its effects be less upon the solids. They may be relaxed or braced, have their sensibility, motion, &c., greatly increased or diminished by different kinds of aliment.

A little attention to these things, says Dr. Buchan, will show how much the preservation of health depends upon proper diet.

Nor is proper diet only necessary for its preservation but also for its restoration. Many diseases may be cured by diet alone, and others greatly mitigated.

It is not my intention, however, at this time to investigate the nature and properties of the various aliments in use among mankind, nor to show their effects upon different constitutions of the human body, but to show some of the more pernicious habits men are liable to run into with regard to their food, quantity they eat, and the influence it produces upon their health.

It is impossible to direct in regard to the quantity the various constitutions, sex and age require. The best rule is to avoid the two extremes. Nature teaches every creature in most cases. When nature calls for food or drink, it is an evidence that nature should be supplied.

But with regard to quality we should act like intelligent men, not like dumb beasts. The food we eat contains frequently the fruits of disease. Unripe or shriveled grain, potatoes defective with spots, and roots or any other vegetables not sound or ripe, is an evidence that they are defective or diseased, and should not be eaten. Grain may be kept too long until musty, and therefore rendered unwholesome. Animal food may also be rendered unwholesome by being kept too long. All animal substances have a natural tendency to putrefaction, and when kept too long are not only rendered repugnant to the taste and smell, but hurtful to the health.

Animals diseased or killed by accident, are also hurtful, as their blood quickly putrifies. The divine injunction given to the Jews, not to eat any creature which had died of itself, has a reference, no doubt, to their health. Animals over-heated are rendered feverish. Butchers should be careful not to overheat their stock before they kill them, as this renders their flesh very unwholesome.

Swine, ducks, and other animals which are grossly fed, are also unwholesome.

The art of cooking is frequently rendered unwholesome by jumbling together half a dozen different ingredients. Pickles, vinegar, salt and spices, are all bad for the health.

Water should also be carefully selected, as it forms the basis of a portion of the solids. That water is most pure which is the most free from foreign bodies. Water takes up parts of most bodies with which it comes in contact, by which means it is often impregnated with metals or minerals of a poisonous nature.

Hard water, or such as is found in limestone countries, is also hurtful. In some countries in which I have practiced, I have found at least one half of the ladies with large goitres upon their necks, and scrofulous tumors as quite prevalent in both sexes. With such water common rosin soap should not be used for washing the body or extremties, as it obstructs the pores, and by that means prevents the escape of the impurities of the body through them. You may saturate the water with saleratus, wood ashes, bran bread, potatoes, or anything of a mucilaginous nature, and this will prevent its deleterious effects, as this is equally necessary with proper diet.

Persons who are weak and relaxed, ought to avoid all viscid food, or such things as are hard of digestion. Their diet, however, ought to be solid, and they should take plenty of exercise in the open air. Such as abound with blood, should be sparing in the use of everything that is highly nourishing, and, in a word, all should study simplicity, and carefully avoid luxuries. Nature delights in the most plain and simple food, and every animal except man follows his dictates. Man alone riots at large, and ransacks the whole of creation in quest of luxuries for his own destruction.

Says Mr. Addison, an elegant writer of the last age : For my part, when I see a fashionable table set out in all its magnificence, I fancy I see gouts, dropsies, fevers, lethargies, with other innumerable diseases, lying in ambuscade among the dishes.

Intemperance in eating is not the only destroyer of our race. How quickly does the immoderate pursuit of carnal pleasures destroy man. Behold that young man who is destroying himself with his own hand, or by debauchery. How soon he begins to complain of headache, glimmering of the eyes, palpitation of the heart, nervousness, pain in the side and back, cold feet, frequently coughs, consumption, paleness, and finally dies a victim, if not corrected by skillful hands, of his own evil conduct, for he is his own murderer. Or if he does not die while young, he has planted the seeds of disease and premature decay in his own body, and must suffer the penalty of a violated law down to the latest hour of his existence.

O could that dear young man know what disease and destruction he is entailing upon himself, when thus indulging in the gratification of carnal pleasure and lust, I believe he would be forever restrained from such a dreadful habit.

The improper use of intoxicating liquors is another destroyer of our race. Every act of intemperance puts nature to the expense of a fever, before she can discharge the poisonous draught. When this is repeated every day, or frequently, it is easy to foresee the consequences. But fevers from intoxication do not always leave the patient in a day; they frequently end in inflammation of the liver, brain, breast, &c.

Dr. Beach, of New York, says: Spirituous liquors inflame the blood, corrode the coats of the stomach, impair digestion, destroy the appetite, and induce many diseases of a fearful character, such as gout, scirrhus of the liver or spleen, dropsy, apoplexy, palsy, madness, and fevers of different kinds.

They also impair the judgment, destroy the memory, and produce intoxication.

Dram drinking produces results terrible in their tendency, and end in the most fearful consequences. One glass makes way for another, and every glass inflames the appetite, stupefies the mind, and renders the man weaker and less capable of resisting the poisonous cup, for poison it most certainly is ; indeed it may not kill as quick, but in the end is certain. Although he may not fall by an acute disease, he seldom if ever escapes those of a chronic character. But drunkenness does not only destroy the miserable creature himself, but the innocent too often feel the dreadful effects of intemperance.

How many wretched orphans are to be seen in almost every town, who once had respectable parents, in affluence and wealth, who are now reduced to mere paupers, and perhaps while borne down with poverty and want, disease as fatal as poison itself, has seized upon them.

How often do we behold the miserable mother, with her helpless infant pining in want, while the cruel father is indulging his insatiate appetite.

Families are not only reduced to want, disease, and death, by intoxication, but even extirpated. Nothing tends more to prevent propagation than intemperance. How often do we find the poor man who labors hard for his scanty morsel, surrounded with numerous offspring, hale and well, while his pampered lord, sunk in ease and luxury, often languishes without an heir. Or if he has any, they are weakly and short lived, and soon the race becomes extinct.

Many resort to this miserable subterfuge in the hour of trouble, for relief. They find a sort of temporary relief, but, alas! " it biteth like a serpent and stingeth like an adder." This solace is short lived, and when it is over, the spirits sink as much below their natural pitch, as strong drink had raised them above it, and hence the necessity of a repetition, and every fresh dose makes way for another, until the unhappy wretch becomes a slave to the bottle, and at length falls a sacrifice to what at first, perhaps, was

taken as medicine or to drown trouble. No
man is as dejected as the drunkard, when
his debauch has gone by. Then he feels his
spirits dejected, disease preying upon him,
nerves unstrung, prostration, poverty, and
want, ten thousand devils haunting him
right and day, and the end a drunkard's
grave and a drunkard's hell.

A want of cleanliness is also injurious to
health; and this is an evil which admits of
no excuse, as God has so plentifully supplied
us with water. The continual discharge
from our bodies through the pores, by perspiration, requires the frequent change of
apparel. Frequent change of apparel greatly promotes the secretion from the skin, so
necessary for health, without which, the matter that should be carried off from the body,
is either retained or re-absorbed from dirty
clothes, and often occasions cutaneous diseases. Itch and several other skin diseases,
are frequently the result of filth. Putrid
and malignant fevers result from the want
of cleanliness. They almost invariably begin among low and filthy inhabitants. They

wear filthy clothes, breathe unwholesome air, live in close, dirty houses, and take but little exercise. Here we find the infection hatched which often spreads far and wide, to the destruction of thousands. Hence cleanliness becomes a matter of public interest; for what will it avail that I keep clean myself, if my neighbor is wallowing in filth and dirt.

Infections are frequently communicated by tainted air. Hence the importance of keeping your cellars, wells, and streets clean. Slaughter houses should also be kept at a proper distance, as nothing renders the air more impure than putrid blood.

Tobacco is another potent enemy to man. It impairs his health. Tobacco is an actual and a virulent poison. Three drops of the oil applied to the tongue of a cat, will destroy its life in from three to seven minutes. Make an incision in a pigeon's leg, and in two minutes it will destroy the action of the limb, and cause violent vomiting.

Kempfer classes it with strong vegetable poisons. Saturate a thread with tobacco oil, and draw it through a wound in an animal, and it will kill quickly. It has been known to do it in seven minutes.

Dr. Maynwaring has asserted in his treatise on scurvy, that tobacco causes scorbutic complaints, and that scurvy has abounded much more since the use of tobacco has become so prevalent, than previous.

Old Mr. Salmon, a man eminent in medical practice, says that snush taking, (meaning snuff,) is productive of great evils ; that it frequently induces apoplexies, and that a hundred has died of this disease to one previous, and most of those deaths I have witnessed from apoplexies, were great snuff takers.

Smoking is also injurious. The saliva which is so copiously drained off by the pipe, is the first and greatest agent Nature employs in digesting food.

Chewing tobacco also destroys the digestion, from the same cause. The unwise custom of chewing and smoking tobacco for many hours in a day, not only injures the salivary glands, producing dryness of the mouth when the drug is not used, but it is supposed that it also produces scirrhus of the pancreas. It also injures the digestion, by throwing off that saliva which the stomach requires, and which the person should swallow.

Again, the practice of tobacco smoking or chewing, saturates the tongue and mouth with tobacco juice, thereby vitiating the saliva that remains, which in this pernicious and poisonous condition, finds its way to the stomach.

In view of these facts, who can wonder, then, that the use of tobacco produces debility, and fixes its deadly grasp upon the organs of vitality, gradually undermining the health and sowing the seeds of disease in the body, which will sooner or later take root and spring up, carrying away its victim to a premature grave. Who can wonder, then, that those who use tobacco, are afflicted with dizziness of the head, faintness, weak stomach, tremulousness, huskiness of the voice, disturbed sleep, nightmare, mental depression, epilepsy, great nervousness, and general debility. It will also produce dyspepsia, and it creates thirst, which is an evidence that saliva is deficient, and hence water is required. Thus you have water instead of saliva for digestion. No wonder, then, you become costive, and many using tobacco are afflicted with piles, &c. Again, the use of tobacco is pernicious, because of its filth. Look at that young man squirting his tobacco juice in bar-rooms, halls, houses of worship, parlors, on fine carpets, ladies' apparel, &c., then add to that the offensive breath of those smoking and chewing, and is it not heart-sickening? Young ladies, would you not turn away from such a young man, or thing, besmeared with the cud and scented with smoke, as you would from the drunkard in his inebriety.

I would therefore exhort boys and young men to flee from these pernicious habits as you would from the bite of an adder. Habits once formed are hard to be corrected, and therefore should be carefully avoided.

PREPARATIONS.

Wine Bitters.

2 drachms Wild Cherry Bark,
2 " Spignant Root,
2 " Solomon Seal Root,
2 " Comfrey Root,
1 " Colombo Root,
1 " Gentian Root,
1 " Chamomile Flowers.

Bruise all, and add boiling water to cover. Let stand one hour; then add 1 quart Domestic or Port Wine.

Dose—½ to 1 wine-glassful three times a day.

This is the best strengthening tonic I have ever found for weak and relaxed systems.

Sweating Powder.

¼ oz. pulv. Opium,
1 " pulv. Gum Camphor,
½ " pulv. Ipecac,
¼ lb. Cream Tartar.
Dose—from ¼ to ½ a teaspoonful.

Mandrake Compound.

Equal parts Pulv. Mandrake Root, **Spear-mint** Herb, and Cream Tartar.

Sassafras Liniment.

½ oz. Sassafras Oil,
¼ " Hemlock Oil,
¼ " Red Cedar Oil,
¼ " Camphor Gum,
¼ " Capsicum,
¼ " Turpentine Spirits,
1 pint Alcohol.

Mix all, and bathe the parts freely and frequently. For pain in the head, back, side, breast, or limbs.

Emetic Preparation.

1 part pulv. Lobelia Seeds,
1 " " " Herbs,
2 " " Ipecac Roots,
1 " " Blood Root.

Mix all thoroughly, and give one teaspoonful every twenty minutes, drinking freely luke-warm water, until the patient vomits freely. This will not only cleanse the stomach, but produce a lively action of the blood, and produce profuse perspiration, and often break up fever without the aid of other medicines, if taken in time.

Cough Drops, for Colds and Coughs.

½ drachm Oil of Almonds,
½ " Balsam of Fir,
½ " Tinct. Balsam Tolu,
½ " Wine,
½ " Tinct. of Black Cohosh.

Mix all, and take from 25 to 30 drops three to five times a day.

Rum Liniment.

Take equal parts of good old Jamaica Rum, Laudanum, and Tinct. of Camphor. For inflammation. Warm and bathe the parts, or apply a cloth wet in the same.

Cayenne Cough Powder.

$\frac{1}{2}$ oz. pulv. Cayenne Pepper,
$\frac{1}{2}$ " " Skunk Cabbage,
$\frac{1}{4}$ " " Wild Turnip,
$\frac{1}{2}$ " " Ipecac,
$\frac{1}{2}$ " " Opium.

Dose.—one eighth of a teaspoonful every four to six hours. For Colds, Coughs, Inflammation of the Lungs, and Difficulty of Breathing.

White Drops.

1 oz. Oil of Sweet Almonds,
1 " Sweet Spirits of Nitre,
1 " Castile Soap, shaved fine,
$\frac{1}{2}$ " Balsam Copaiba,
$\frac{1}{4}$ " Spirits of Turpentine,
1 drachm Camphor Gum.

Sudorific Drops.

2 oz. of Ipecac,
2 "   Saffron,
2 "   Camphor Gum,
2 "   Virginia Snake Root,
2 "   Opium,
2 "   Motherwort Extract,
3 quarts Holland Gin.
Mix all, and let stand two weeks. Strain or filter.

Dose—1 tea-spoonful given in a cup of Catnip tea every hour or two, until it produces perspiration.

Cholera Compound.

1 oz. Tinct. Camphor,
1 "   "   Rhubarb,
1 "   "   Opium.
Mix all.
Dose—from 20 to 30 drops every fifteen minutes.
For Dysentery, Cramp of the Stomach, Pain of the Bowels, Billious Cholic. &c.

Compound Tincture of Spearmint.

Take the best old Holland Gin and apply to the green herb. Bruise and press out the juice, and add equal parts Sweet Spirits of Nitre and Tincture of Blue Flag.

Dose must be regulated by the urgency of the case as well as the habits of the patient, say from 1 tea-spoonful to 1 table-spoonful every thirty to sixty minutes.

For Stoppage of Urine and Disease of the Kidneys and Prostate Glands, &c.

Red Ointment.

3 oz. of Fresh Butter,
2½ drachms of Red Precipitate,
1   "   Prepared Tutty,
1   "   . Camphor Gum, **dissolved in** Olive Oil.
½ oz. of white Wax.

Melt the Wax in the Oil, and while cooling, stir in the other ingredients, and continue to stir until cold.

For Sore Eyes, Eruptions, &c.

Neutralizing Powder.

1 oz. pulv. Turkey Rhubarb,
1 "   "   Saleratus,
1 "   "   Peppermint Herb.
Dose—1 large tea-spoonful.

Add 1 pint of boiling Water. When cold, strain, sweeten, and add 1 table-spoonful of brandy, and take from 1 to 2 table-spoonsful every thirty to sixty minutes.

No. 4 Powder.

½ oz. pulv. Dandelion Root,,
½ "   "   Mandrake Root,
¼ "   "   Blood Root.

Mix all, and add a few drops of Peppermint Oil, and mix well.

Dose—every morning the amount that will lay on a five cent piece.

Camphor Compound.

Take Whisky, Camphor and Water, warm wet cloths and apply to the parts inflamed.

Sarsaparilla Syrup.

1 lb. American Sarsaparilla,
½ " Guaicum Shavings,
¼ " Elder Blows,
¼ " Burdock Root,
¼ " Bark of Sassafras Root.

Add water ; boil well ; turn off ; add more ; boil and turn off ; and so continue until the roots are boiled soft ; then strain and simmer down to one gallon. Add a little spirits, and sweeten to taste.

For impurities of the blood.

Dose—from a half to one wine-glassful, three times a day.

Caustic Potash.

You will take Hickory, hard Maple, or white Elm, and burn into ashes ; then leach and strain the lye free from ashes, and simmer down until dry ; then stir until cold and fine.

Fire Oil Ointment.

Take 1 oz. of pure Fire Weed Oil and 1 lb. of hog's Lard. Mix all cold and it is fit for use.

Gargle.

To cure common Sore Throat, take 1 Sumach Bob and add 1 pint of boiling water, and gargle the throat three to five times a day.

To Remove Warts and Corns.

Take 1 lb. of common Potash, and half a pint of water ; add half an oz. of Extract of Belladonna, 1 oz. of Gum Arabic, and a little Wheat Flour, so as to form a paste. Keep well corked. Apply a little to the parts affected, and let it remain about five minutes. Then loosen the edges with a sharp knife, and presently you can take it out by the roots. Then add a little Sweet Oil and Vinegar, and keep the parts warm and dry until healed.

Strengthening Plaster.

Take 3 parts Hemlock Gum,
"   1   "   White Pine Gum.

Melt and strain. Spread on a thin piece of leather, and apply while moderately warm.

Hair Dye.

Preparation No. 1.

Take 2 drachms Gallic Acid.
"   1 oz. Alcohol,
"   3 oz. soft Water.

Mix all.

After cleansing the hair well, and drying, apply it with a tooth-brush. Let it dry well.

Preparation No. 2.

Take 2 drachms Nitrate of Silver,
"   1   "   Spirits Ammonia F.F.F.F.
"   1 oz. soft Water.

Mix.

After applying No. 1 as directed, then apply No. 2 in like manner.

Katharion.

To restore Hair in the head ; also to give it gloss and beauty.

Take 1 pint of Alcohol,
"   1 oz. Cantharides Spirits,
"   4 oz. Castor Oil,
"   ½ oz. Bergamot.

Pain Extracting Plaster.

For Colds, Coughs, and Spinal Affections.
Take transparent Burgundy Pitch, and add Beeswax for consistency.   Melt, spread, and apply.

Anti-Spasmodic Drops.

½ oz. Fluid Extract of Ladies' Slipper.
½ "   "   "   Catnip,
½ "   "   "   Scull Cap.
Mix all.
Dose—from 5 to 15 drops once in two hours.

This is a valuable medicine for Headache, Neuralgia, Nervousness, &c.

Fire Ointment.

Pulverized Charcoal mixed with hog's Lard, spread on cloth, and applied, will extract the fire in a few minutes and relieve the pain.

Balm of Gilead Ointment.

Take the Balm Buds and the inner Bark of Sweet Elder, equal parts. Add fresh Butter to cover. Simmer slowly four or five hours, until crisped. Then press out, and when cold it is fit for use.

This is a valuable medicine for Cuts, Burns, Scalds, Frosted Limbs, Ulcers, &c. It may be used on any eruption where the skin is broken.

White Liniment.

For Chilblains, Rheumatism, Sprains, and Bruises, on man or beast.

Take 1 pint of Olive Oil and 8 pints of Ammonia—Hartshorne. Mix and shake well before using. Keep well corked, and apply freely and frequently.

Soothing Syrup.

Take Motherwort Herb, 4 parts, and Poppy Blows, 1 part ; extract the substance, sweeten well, and give the child a few drops according to age and strength.

Worm Powder.

Take ½ oz. of Senna,
" ½ " Carolina Pink,
" ½ " Manna.

Add 1 quart of boiling water ; let stand six hours ; strain and sweeten, and give the child half a tea-spoonful three or four times a day, say a child six years old. Always vary according to age.

Inflammatory Liniment.

Take equal parts good old Jamaica Rum, Laudanum, and Tinct. of Camphor. Warm, and bathe the parts freely and frequently.

Diptheria or Putrid Sore Throat.

When first attacked, you will give the patient a vegetable Emetic ; immerse the feet in warm weak Lye ; bathe the throat with the rum liniment. Or you may take bitter Herbs, such as Catnip, Wormwood, Hops, Tansy, Smartweed, or any other bitter herbs. Boil. Steam the neck and head. You may also let the patient inhale the steam of the same from a teapot, or any other vessel most convenient, and give the patient Cathartics freely. Above all, keep the stomach and bowels well cleansed ; But should the patient be attacked violently, and liable to choke with the mucus collecting in the throat, take a quill cut at each end, fill it with pulverized Bloodroot, and blow it rapidly into the throat, and repeat if necessary. This remedy has never failed. It will lessen the inflammation and cause the patient to throw off the mucus. You may also bind warm herbs about the neck, as warm as can be borne, and change them frequently.

Another Remedy for Felon.

Take chamber Lye and Copperas; place the mixture over the fire and heat it; then hold the affected part in it as long as you can bear it; cool a little and apply again, until the part suppurates, which will be in a short time, after which make a poultice of Slippery Elm and weak Lye, and apply until thoroughly cleansed; then apply a common healing Salve.

Diarrhea.

Take 1 tea-spoonful pulv. Soda,
" 1 " " Rhubarb,
" 2 " Peppermint Herb.
Mix all. Then to 1 tea-spoonful add half a tea-cupful of boiling water; sweeten well with loaf sugar; you may also add a little brandy.

Dose for an adult, 2 table-spoonsful every four hours.

This seldom fails to cure.

Frozen Feet, Chilblains, &c.

Take 2 oz. Organum Oil,
" 2 " Spirits Turpentine,
" 2 " Hartshorne,
" ½ pint good Brandy,
" ½ " Vinegar.

Mix all, and bathe freely and frequently.
This is also good for pain in the back, &c.

Bloody Flux.

Take an apple, cut out the core and fill it with honey, wrap it in a cloth; then roast it and feed it to the child.

CONTENTS.

Ague and Fever, or Intermittent Fever............ 7
Remittent Fever................................. 10
Inflammatory Fever 11
Continued Fever................................ 12
Scarlet Fever.................................. 14
Infantile Fever................................ 16
INFLAMMATORY DISEASES........................ 17
Inflammation of the Brain..................... 13
   "   "   Ear....................... 19
   "   "   Lungs 25
   "   "   Stomach.................. 28
   "   "   Womb.................... 29
   "   "   Bladder.................. 30
   "   "   Eye...................... 50
   "   or Ague of the Breast............. 47
Mumps... 19
Quinsy.. 20
Croup 22
Hooping Cough................................. 23
Colds and Coughs.............................. 24
Pleurisy...................................... 26

118 CONTENTS.

Small Pox.. 31

Measles... 35

Delirium Tremens............................... 36

Cholera Morbus.................................. 38

Asiatic Cholera................................. 40

Cholera in Children............................. 40

Vomiting.. 42

Water Brash..................................... 43

Cramp in the Stomach............................ 44

Heartburn....................................... 44

Hiccough.. 45

Bleeding of the Nose................. 45

Jaundice.. 46

Locked-Jaw 49

Felon.................................51, 69, 115

Chilblains............................52, 116

Boils... 52

Ague in the Face or Jaw......................... 53

Inverted Toe Nail 53

Corns and Warts................................. 54

Scalds and Burns................................ 54

Ringworm.. 55

Itch.. 56

Sprains 57

Stye of the Eye................................. 68

To preserve the Eye and restore from partial Blindness.. 69

Diptheria or Putrid Sore Throat...............114

Diarrhea......................................115

Frozen Feet...................................116

Bloody Flux...................................116

Antidote for Poisons.................... 58
Ivy Poison... 61
Bite of Snake.. 62
Bite of Mad Dog.................................... 63
Diseases of Children............................. 65
Purging or Looseness............................. 66
Teething... 66
Sores about the Ears, Groins, &c................. 66
Sore Mouth.. 67
Convulsions from Teething........................ 67
Soreness about the Navel.......................... 67
Rupture... 67
Tongue Tied... 68
Worms.. 68

Lecture on the Nervous System, and Diseases
    incident thereto......................70—101

PREPARATIONS.

Wine Bitters... 101
Sweating Powder...... 102
Mandrake Compound............................. 102
Sassafras Liniment................................. 102
Emetic Preparation................................ 103
Cough Drops, for Colds and Coughs............. 103
Rum Liniment....................................... 104
Cayenne Cough Powder........................... 104
White Drops... 104
Sudorific Drops..................................... 105
Cholera Compound................................. 105
Comp. Tinct. Spearmint........................... 106

Red Ointment.................................. 106
Neutralizing Powder.......................... 107
No. 4 Powder................................. 107
Camphor Compound............................ 107
Sarsaparilla Syrup........................... 108
Caustic Potash............................... 108
Fire Oil Ointment............................ 109
Gargle....................................... 109
To Remove Warts and Corns.................... 109
Strengthening Plaster........................ 110
Hair Dye..................................... 110
Katharion.................................... 111
Pain Extracting Plaster 111
Anti-Spasmodic Drops......................... 111
Fire Ointment................................ 112
Balm of Gilead Ointment...................... 112
White Liniment............................... 112
Soothing Syrup 113
Worm Powder.................................. 113
Inflammatory Liniment........................ 118

```
* 9 7 8 3 3 3 7 2 9 6 9 9 5 *
```